Calcium

John Farndon

BENCHMARK BOOKS

MARSHALL CAVENDISH
NEW YORK

Benchmark Books
Marshall Cavendish Corporation
99 White Plains Road
Tarrytown, New York 10591-9001

www.marshallcavendish.us

Library of Congress Cataloging-in-Publication Data
Farndon, John.
Calcium / John Farndon.
p.   cm. — (The elements)
Includes index.
Summary: Describes the origins, properties, chemical activity, and
uses of the element calcium.
ISBN 0-7614-0888-6
1. Calcium—Juvenile literature. [1. Calcium.] I. Title.
II. Series: Elements (Benchmark Books)
QD181.C2F37 1999
546'.393—dc21   98-55094 CIP AC

Printed in China

Picture credits
Corbis (UK) Ltd: 6, 8, 9, 12–13, 14, 20, 23, 27, 30.
Image Bank: 25, 26.
Image Select: 15, 16, 19.
John Bates: 10, 11, 21.
Mary Evans Picture Library: 18.
Paul Crawford: 4.
Science Photo Library: 12, 22, 24.

Series created by Brown Packaging Partworks
Designed by wda

Contents

What is calcium?

Calcium is a soft, silvery-white metal, but it does not occur naturally in this pure form. However, because it forms compounds with many other elements, it is the fifth most abundant element in Earth's crust, and its compounds are among the most important substances on Earth.

Most calcium compounds are white solids, often called limes. Whenever you see a white solid, the chances are that it is a calcium compound. Among the most familar calcium compounds are drawing chalks, white chalk cliffs, porcelain, teeth, cement, seashells, plaster casts on broken limbs, and the white scale (limescale) found on faucets and inside kettles.

Without calcium, life would be impossible: it is essential for the growth of both plants and animals. Calcium plays many roles in your body. It adds rigidity and strength to bones and teeth, helps to control muscles, and aids digestion. Many people get the calcium the body needs from milk, cheese, and other dairy foods.

These tall limestone cliffs are in Bryce Canyon, Utah. Pure limestone—a compound of calcium—is white, but traces of other elements, such as iron, have given these rocks their rich reddish coloration.

The calcium atom

Like all elements, calcium is made up of tiny particles called atoms. Calcium has an atomic number of 20, which means that the nucleus at the center of each calcium atom contains 20 even smaller particles called protons. Protons have a positive electrical charge. Circling the nucleus—in several layers called shells—are 20 negatively charged electrons.

The nucleus of most calcium atoms contains 20 neutrons as well as 20 protons, giving it an atomic mass of 40. Some calcium atoms have as many as 28 neutrons, and these variations, or isotopes, have a greater atomic mass—up to 47.95.

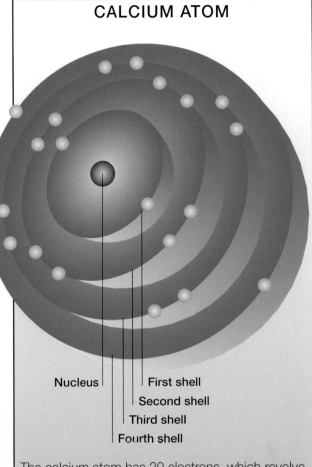

CALCIUM ATOM

Nucleus | First shell
Second shell
Third shell
Fourth shell

The calcium atom has 20 electrons, which revolve around the nucleus in four shells. There are two electrons in the inner shell, eight in both second and third shells, and two in the outer shell.

DID YOU KNOW?

ALKALINE-EARTH METALS

In the 18th century, chemists called any substance that was insoluble in water and unaffected by heat an "earth." At the time, they knew none of the alkaline-earth metals as elements, but they knew two compounds common in nature—calcium oxide and magnesium oxide. They called these alkaline earths, because they were not only almost insoluble and little affected by heat, but they were also alkaline (opposite to acidic).

Later, when British chemist Humphry Davy split these compounds to isolate, for the first time, the pure elements calcium and magnesium, they were named alkaline-earth metals. This name was also applied to the other four elements of the group when they were finally discovered.

The periodic table

Calcium, along with five other similar elements, including magnesium, is known as an alkaline-earth metal. Together, these six elements make up Group II of the periodic table of elements. Each of these alkaline-earth metals has two electrons in the outermost shell of its atom, and this explains why they all have similar chemical properties.

Special characteristics

COMPOUND FACTS

The list of common calcium compounds is a long one, and the following are just a few of those you might come across:

⬤ calcium carbonate—much used in the building industry and in steelmaking and to make sodium carbonate, which itself has many industrial uses

⬤ calcium chloride—sometimes spread on dry roads to reduce dust

⬤ calcium cyanamide—used for making drugs and plastics

⬤ calcium cyanide—used in goldplating

⬤ calcium cyclamate—a white powder once used as a sweetener in drinks

⬤ calcium hydrogen phosphate and calcium nitrate—used as fertilizers

⬤ calcium hydroxide—used to make building mortar

⬤ calcium hypochlorite—used as a bleach

⬤ calcium oxide—used in making porcelain, glass, cements, and fertilizers

⬤ calcium phosphate—used to make phosphoric acid, which is in turn made into fertilizers

⬤ calcium pyrophosphate—used in toothpaste

⬤ calcium silicate—used in glassmaking

⬤ calcium sulfate—used to make plaster of paris and plaster wallboards.

Calcium does not occur naturally in pure form because it reacts so easily with other elements. In water, for instance, it reacts to form a solution of the compound calcium hydroxide (known as limewater) and the gas hydrogen.

However, chemists have been able to separate calcium artificially from its compounds. They found a soft metal (you can mark it with a fingernail) that burns with a red flame. It has a melting point of 1,542°F (839°C), a boiling point of 2,700°F (1,484°C), and a relative density of 1.55 grams per cubic centimeter, which means that a given volume of calcium would weigh 55 percent more than the same volume of water.

Calcium is usually stored in the form of tiny pellets that look rather like peanuts. These pellets of pure calcium are made by

This is a sample of pure calcium metal. It looks gray because, through exposure to the air, it has become coated with a thin film of calcium oxide.

the electrolysis of molten calcium chloride. When fresh, the pellets are silvery white, but they quickly tarnish to gray as the calcium reacts with oxygen in the air to form a coating of calcium oxide. If the air is moist, the calcium oxide in turn quickly reacts with the water, changing to calcium hydroxide (see page 17).

Compound interest

Because it is so reactive, calcium forms a wide range of common compounds. Many of these compounds form white crystals that build up into rigid white masses, as in bones and rock. When powdered and mixed with water, many compounds of calcium can feel slimy. This is why many calcium compounds have the word "lime" in their name (from *limus*, Latin for *slime*).

Calcium oxide is known as quicklime. It gets the name "quick" (an old English word for "living") because when water is dripped on it, it seems to contort and swell (see page 17). Calcium hydroxide is known as "slaked lime"—perhaps because it is used in the garden to "slake" the plant's "thirst" for lime in acid soils (see page 24).

When the water is squeezed or evaporated out of calcium compounds, the slime can turn to hard cement. In the case of calcium carbonate, hard rocks such as chalk and limestone are formed. By contrast, calcium sulfate (called gypsum) forms a much softer plaster.

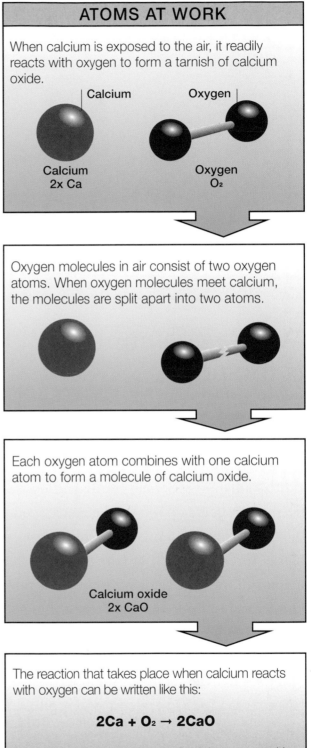

ATOMS AT WORK

When calcium is exposed to the air, it readily reacts with oxygen to form a tarnish of calcium oxide.

Calcium

Oxygen

Calcium
2x Ca

Oxygen
O_2

Oxygen molecules in air consist of two oxygen atoms. When oxygen molecules meet calcium, the molecules are split apart into two atoms.

Each oxygen atom combines with one calcium atom to form a molecule of calcium oxide.

Calcium oxide
2x CaO

The reaction that takes place when calcium reacts with oxygen can be written like this:

$2Ca + O_2 \rightarrow 2CaO$

This shows that two atoms of calcium react with a single molecule of oxygen to form two molecules of calcium oxide.

Where calcium is found

Clearly preserved in this piece of limestone rock found in Germany are the fossilized remains of a prehistoric crayfish.

Calcium is found in small quantities throughout the universe. Most of Earth's calcium probably formed at least five billion years ago in the heart of a red giant star. The calcium was formed from alpha particles, which are the nuclei of helium atoms. The enormous pressures and temperatures inside the star caused these particles to fuse together to make other elements—first, carbon and oxygen atoms, and then heavier elements such as calcium, magnesium, silicon, and argon.

In the beginning

Earth formed from the remains of such a giant star. As the hot young Earth slowly cooled, heavy elements such as iron sank to the core, but lighter elements such as silicon and oxygen floated up to form the solid crust. Because calcium readily forms compounds with silicon and oxygen, it too was carried up into the crust. The result is that calcium is one of the most dominant elements in Earth's crust. Although it is not found pure in nature, in compound form calcium makes up some 3.5 percent of Earth's crust.

Crust funds

The bulk of the calcium compounds in Earth's crust are found in rocks, in the form of the mineral calcite. Calcite is

8

essentially crystals of calcium carbonate. Chalk, marble, and most of the different kinds of limestone (except dolomite) are all made mainly of calcite.

Most of these rocks were originally sediments that had settled on the seabed. Over millions of years, the sediments were squeezed by other sediments piling on top of them, so they gradually became transformed into rock.

Much of the calcite on Earth formed chemically. Earth's early atmosphere was very rich in carbon dioxide, which billowed out from the many volcanoes bursting through Earth's newly formed crust. Some of the carbon dioxide stayed in the air, but much of it dissolved into the water of the oceans. Calcium that was dissolved in the oceans combined with the dissolved carbon dioxide to form solid calcite. The calcite sank to the bottom and was deposited on the seabed.

Some of the calcite in rocks, however, was formed organically—that is, by living creatures. Calcium compounds play an important part in many living things (see pages 26–27), in bones and shells, for example. Fragments of the shells, bones, and other bits of billions of sea creatures piled up on the sea floor, adding huge quantities of organic calcite to the calcite formed chemically.

This is a sample of the mineral calcite, which is made up of molecules of calcium carbonate that have arranged themselves into crystals.

Calcium in the landscape

Most limestones are a mix of chemical and organic calcium carbonate. A few, such as shelly and reef limestones, and chalk, are made almost entirely of fossil sea creatures, altered by chemical changes over time into calcite.

The relative hardness of limestone makes it a useful building material. For example, it is quarried and converted, by heating, into calcium oxide. This is made into cement, which is the basis of concrete.

Chalk, by contrast, is a soft white rock that is not suitable for building. It is almost pure calcite. In the Cretaceous period (100 million years ago, when dinosaurs roamed the Earth) billions of tiny algae-like plants called coccoliths grew across the seabed. These plants, along with the shells of tiny sea creatures called foraminifera, gradually hardened after they died and turned to calcite. Over time, the calcite built up into layers of chalk. The chalks once used in schools came from rock formed this way.

These cliffs in northern France are made of chalk, a soft form of calcium carbonate made millions of years ago from dead plants and tiny animals.

Oolites and dolomite

Oolitic limestones are very rich in fossils that contain tiny balls of calcite called ooliths, after the Ancient Greek word *oon*, which means "egg." Some ooliths look like fish eggs (called roe), and oolitic limestone can be known as roe limestone. Ooliths were made from grains of silt that became coated in lime as they rolled about in lime-rich mud on the seabed.

DID YOU KNOW?

HOW FOSSILS FORM

When creatures such as shellfish die and fall to the sea floor, their soft body parts soon rot away. But the hard shells, made mainly of the mineral aragonite, which contains calcium carbonate, may be buried intact by sediment (material, such as sand, deposited on the sea floor). Over millions of years, the shells dissolve as water trickles through the sediments. But the dissolved aragonite may recrystallize within the cavity, like jelly in a mold, leaving a perfect replica or "fossil" of the shell in calcite.

The facade of this cathedral in Florence, Italy, is covered with marble of several different colors.

Dolomite looks much like other limestones but it is chemically different. Most limestones are essentially calcite—calcium carbonate. But dolomite is calcium magnesium carbonate. Dolomite probably started off as mostly calcite, but it is thought that seawater must have seeped into the rock and replaced some of the original calcium with magnesium.

Marble

Marble is another rock that began as ordinary limestone. It is still essentially calcite, but the crystals have been

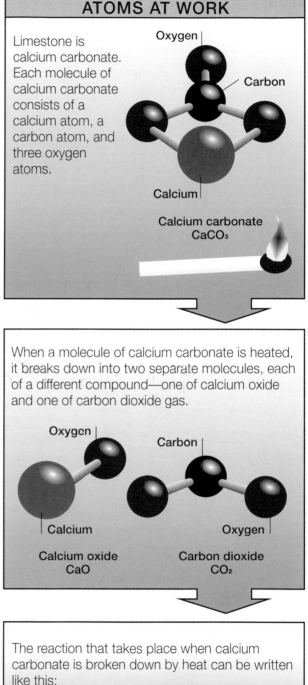

ATOMS AT WORK

Limestone is calcium carbonate. Each molecule of calcium carbonate consists of a calcium atom, a carbon atom, and three oxygen atoms.

Oxygen

Carbon

Calcium

Calcium carbonate
$CaCO_3$

When a molecule of calcium carbonate is heated, it breaks down into two separate molecules, each of a different compound—one of calcium oxide and one of carbon dioxide gas.

Oxygen

Carbon

Calcium

Oxygen

Calcium oxide
CaO

Carbon dioxide
CO_2

The reaction that takes place when calcium carbonate is broken down by heat can be written like this:

$$CaCO_3 \rightarrow CaO + CO_2$$

This shows that one molecule of calcium carbonate breaks down into a molecule of calcium oxide and a molecule of carbon dioxide.

completely changed through contact with hot magma underground. The heat cooks the rock so that the calcite is recrystallized to form a beautiful shiny white rock. Very pure limestone produces brilliant white marble. Limestones with impurities can give the typical marble colored streaks, as the impurities are changed by the heat to brightly colored minerals. Iron, silica, and magnesium, for example, change to green olivine or amber-colored garnet.

Calcite and other crystals

Minerals are solid chemicals found naturally in the ground. Calcium is one of the most common elements found in minerals. The most widespread of these calcium-rich minerals is calcite, a form of calcium carbonate. Calcite is the main ingredient of limestone rocks. Under some circumstances, it forms large crystals in

many shapes, such as "dogtooth spar," which looks like the fangs of a dog. Iceland spar is a type of calcite that forms in clear, flat crystals like pieces of ice. If you look through a crystal of Iceland spar, you see things behind as a double image. Most calcite crystals are white, but they can be tinted many colors by impurities.

Calcium carbonate also forms another mineral, aragonite. Aragonite is the mineral that makes up seashells, but it is much rarer than calcite in rocks. Typically it forms in crusts around hot springs.

Gypsum, shown here in an attractive variety called a desert rose, is a mineral form of calcium sulfate. It is widely quarried and used to make plaster.

At Mammoth Hot Springs in Yellowstone National Park, Wyoming, calcite (in the form travertine) has been deposited in a series of terraced pools. The pools formed as very hot water under pressure reached the surface from underground thermal springs. As the water cooled, minerals dissolved in it were deposited.

Calcium minerals such as aragonite and calcite are highly soluble in acid, and geologists often drop them in hydrochloric acid to identify them. If the mineral sample fizzes (as carbon dioxide gas is given off), it is likely to be either calcite or aragonite.

Other calcium-rich minerals are gypsum (calcium sulfate), fluorite (calcium fluorite), and apatite (a form of calcium phosphate). Gypsum is the source of plaster of paris, used

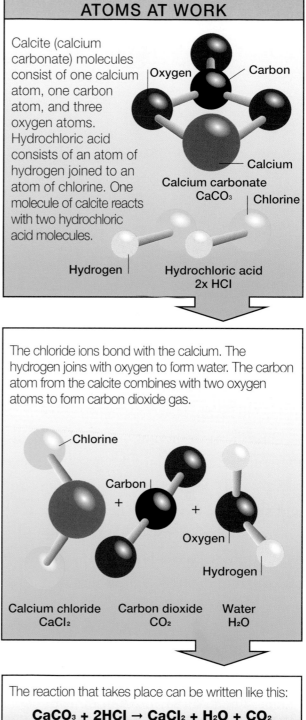

ATOMS AT WORK

Calcite (calcium carbonate) molecules consist of one calcium atom, one carbon atom, and three oxygen atoms. Hydrochloric acid consists of an atom of hydrogen joined to an atom of chlorine. One molecule of calcite reacts with two hydrochloric acid molecules.

Oxygen | Carbon

Calcium

Calcium carbonate
$CaCO_3$ | Chlorine

Hydrogen | Hydrochloric acid
2x HCl

The chloride ions bond with the calcium. The hydrogen joins with oxygen to form water. The carbon atom from the calcite combines with two oxygen atoms to form carbon dioxide gas.

Chlorine

Carbon |
+

Oxygen |

Hydrogen |

Calcium chloride
$CaCl_2$ Carbon dioxide
CO_2 Water
H_2O

The reaction that takes place can be written like this:

$$CaCO_3 + 2HCl \rightarrow CaCl_2 + H_2O + CO_2$$

This shows that one molecule of calcium carbonate reacts with two molecules of hydrochloric acid to form one molecule each of calcium chloride, water, and carbon dioxide.

These strange tooth-like deposits of calcite are called stalactites and stalagmites. They form inside limestone caves from dripping water that contains dissolved calcite.

on walls and ceilings. Huge quantities of gypsum are quarried the world over. In the United States, it is quarried in California, Iowa, Michigan, New York, and Texas.

Fluorspar is a kind of fluorite that sometimes forms beautiful blue and yellow crystals of Blue John (from the French *bleu* and *jaune* for "blue" and "yellow").

Apatite mixes with soil to form the main source of phosphorus for plants. It is also the main constituent of human and animal bones.

Ups and downs of calcium deposits

Calcium-rich rocks not only form on the seabed. They can form when mineral waters evaporate, leaving solids behind. Tufa or dripstone, for example, forms where water enriched with calcite, which has dissolved from limestone, drips from the walls and ceilings of caverns. The drips form into long, hanging icicle-like shapes, called stalactites. Where stalactites continue to drip on to cave floors, they gradually build up to form pinnacles, called stalagmites.

These drips of calcite can also build up into fantastic formations around pools. Around hot springs, a crust of calcite—in a form called travertine (see page 13)—can build up as the water evaporates.

Travertine is a pale honey color, often with attractive bands of different tone. Many of the buildings in Ancient Rome were made from travertine.

How calcium was discovered

In the 17th century, English scientist Robert Boyle (1627–1691) suggested that there are such things as "elements," basic substances that can be combined to make compounds. By the time of the French Revolution in the 1790s, some 30 of the 100 or so elements we know today had been identified. These included oxygen and nitrogen, first isolated by French scientist Laurent Lavoisier (1743–1794), the greatest chemist of his time. Other elements recognized by Lavoisier included metals such as mercury, tin, iron, lead, copper, silver, and gold; nonmetals such as carbon and phosphorus; and gases such as hydrogen.

But there were other substances such as lime (now known to be the compound calcium oxide) and magnesia (magnesium oxide) that Lavoisier considered to be elements, just because they could not be broken down by any methods then known.

In 1807, English chemist Humphry Davy (1778–1829) began experimenting with electricity, revealed in 1799 with the

SIR H. DAVY.

The English chemist Humphry Davy is celebrated on this cigarette card together with his best-known invention—the Davy lamp.

invention by Alessandro Volta (1745–1827) of the voltaic pile (a kind of battery). By connecting the electrodes of a voltaic pile to a mixture of mercury and lime (calcium oxide), Davy was able to separate the lime into oxygen and a new element, which he called calcium.

How calcium reacts

Calcium is one of the most reactive of all metals. Calcium atoms have only two electrons in the outer shell, and these are readily lost to form positively charged particles called ions.

Many calcium compounds are ionic—that is, compounds formed by a bond between oppositely charged ions. Since calcium forms positive ions, it joins with negative ions, such as chloride, carbonate, and sulfate ions to form calcium chloride, calcium carbonate, and calcium sulfate. These compounds, formed when an acid reacts with a base, are called "salts." Like most ionic compounds, these calcium salts are solid crystals with high melting points.

They are less soluble in water than salts of metals in Group I of the periodic table, but they react strongly with acids.

Compounds with oxygen and hydrogen

Calcium oxides and hydroxides are quite strong bases, which is why slaked lime (calcium hydroxide) is used in the soil to neutralize acidity, as well as to make bleach. Because it reacts violently with moisture (see page 17), calcium oxide must be turned into calcium hydroxide before it can be used on the soil.

This 19th-century illustration shows the bodies of victims of the London Plague of 1665 being put in a mass grave. The infected bodies were covered with quicklime to speed up their decay.

Calcium oxide

Calcium oxide (quicklime) is a widely used calcium compound—used to neutralize soil, and in cement and concrete. It is not easily available in nature but can be made from the calcium carbonate in limestone. When heated, limestone breaks down into calcium oxide and carbon dioxide gas.

Calcium hydroxide

Calcium oxide is called quicklime because it reacts violently with water. In moist air, calcium oxide quickly turns to calcium hydroxide, which forms a white crust. If water is poured on crystals of calcium oxide, the result is spectacular. The water soaks into the crystals and seems to vanish, but suddenly the crystals swell rapidly, heat up, and give off steam as the water boils.

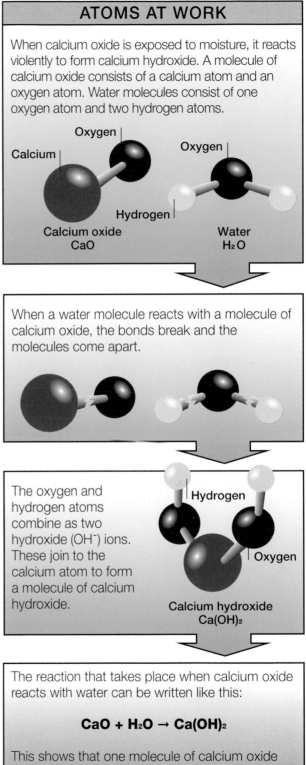

ATOMS AT WORK

When calcium oxide is exposed to moisture, it reacts violently to form calcium hydroxide. A molecule of calcium oxide consists of a calcium atom and an oxygen atom. Water molecules consist of one oxygen atom and two hydrogen atoms.

Oxygen

Calcium

Oxygen

Hydrogen

Calcium oxide
CaO

Water
H_2O

When a water molecule reacts with a molecule of calcium oxide, the bonds break and the molecules come apart.

The oxygen and hydrogen atoms combine as two hydroxide (OH^-) ions. These join to the calcium atom to form a molecule of calcium hydroxide.

Hydrogen

Oxygen

Calcium hydroxide
$Ca(OH)_2$

The reaction that takes place when calcium oxide reacts with water can be written like this:

$$CaO + H_2O \rightarrow Ca(OH)_2$$

This shows that one molecule of calcium oxide reacts with one molecule of water to form a single molecule of calcium hydroxide.

In the limelight

We take bright lights in theaters and cinemas so much for granted that it is hard to imagine that until the late 18th century, people had to watch stage plays by flickering candlelight. Things improved with the invention of the gas lamp in the 1790s, but only in 1822 did theaters get a really bright light, when English physician Goldsworth Gurney (1793–1875), created an amazingly bright, steady light by applying a mixture of oxygen and hydrogen to pellets of calcium.

At first, this light was called the Drummond Light, because it was used by Thomas Drummond (1797–1840) of the Royal Engineers to provide lighthouses with a bright beam for the first time. Its possibilities were soon realized by the theater, where "limelight" boxes were set up at the front of the stage to illuminate the actors on stage with an intensity that had never been seen before. Many actors gravitated towards the front of the stage, so that they could be brilliantly lit— hence the phrase "in the limelight."

In London, England, in 1832, British scientists Cooper and Carry developed a way of using limelight to make very bright microscope images—helping microbiology to make great progress. Then, in 1839, Frenchman Louis Daguerre (1789–1851) pioneered the first practical photographic process. Daguerre tried using limelight to take "flash" photographs, but it provided too harsh a light.

Moving pictures

Following Daguerre's experiments, French scientists Donné and Foucault adapted Cooper and Carry's device into a machine for projecting photographs. In 1844, in Paris, France, they presented the first photographic show, using large glass slides.

For the next 30 years, projectors were made with either limelight or with the

A so-called "limelight man" operating his equipment at the turn of the century, a common sight in theaters at the time.

electric arc lamp, invented by Humphry Davy (page 15). Both were very bright, but hard to use. With limelight, the oxygen and hydrogen had to be prepared first. Arc lamps required huge banks of galvanic batteries, so home users continued to use the dim light of domestic oil-lamps.

Few of the first experiments in making moving pictures used limelight projectors, yet this type of early projector stayed in use for public events until the 1890s, when electricity supplies from power stations began to become commonplace.

Limelight and a mirror are used to persuade the people in this 19th-century theater audience that they are seeing a ghost walk across the stage.

Calcium in building materials

Limestone is a popular and attractive building material. After being quarried, this limestone has been shaped into neat blocks.

Of all building materials, compounds of calcium are the most important. Most cities could not exist without them.

Limestones have been used as building stone since the days of Ancient Egypt. Most limestones are fairly hard, yet easily cut and worked. Everything from the pyramids of Ancient Egypt and famous buildings such as St. Paul's Cathedral in London and Chartres Cathedral in France, to humble stone huts and walls are built from limestone cut from quarries.

Sometimes the stone is cut into neat blocks and smoothed. Sometimes it is used as rough chunks. Limestone buildings can usually be recognized by their attractive pale gray color, often mottled with lichen.

As soon as it is exposed, limestone is corroded by the acid in air, which is why old limestone buildings look weathered.

Marble is also widely used as a building material—to create beautiful floors and pillars and to clad in brilliant white such buildings as the Capitol in Washington, D.C., and the Taj Mahal in India.

Manufactured building materials

Although many buildings are still built from limestone, it is as mortar, cement, and concrete that calcium compounds are now most widely used in construction. Hardly a road, wall, apartment block, or office building in the world is put up without the use of one of these three compounds.

Mortar

Mortar, made from quicklime, is an ancient discovery. For thousands of years, it has been used to "glue" bricks and stones together. When quicklime is mixed with water, it turns to sticky slaked lime. As the water evaporates, the slaked-lime mortar hardens, gluing the stones together.

Concrete and cement

Basic quicklime mortar tends to crumble, but a tougher, longer-lasting mortar is made by adding cement (a powdered mix of calcium compounds) to sand and water. Cement was discovered 2,000 years ago by the Romans, who were even able to make it set underwater, but it was forgotten until rediscovered by British engineer John Smeaton (1724–92) in the 1760s. Cement can be made much tougher by adding gravel, to turn it into concrete, or steel bars to create reinforced concrete. When freshly made and wet, concrete can be poured like a fluid into any shape, but it soon dries out and turns rock hard.

Portland cement is the cement that is most widely used, but all cements are made from a mix of limestone and clays. The mix is made into a mud-like "slurry" and fed into a rotating cylinder, or kiln. When the mix is heated to 2,460°F (1,350°C), the calcium carbonate and clay react to form lumps, which are crushed to a fine gray powder when cool.

The basic ingredient of cement is calcium oxide, made as calcium carbonate reacts to heat in the kiln. Then the calcium oxide reacts with oxides of aluminum and silicon in the clay to give a mix of calcium silicates and aluminates. When water is added to these compounds—in the form of cement powder—a mix of hydrates, and calcium aluminosilicates is produced, and the reactions cause the cement to set hard.

An infamous use of concrete was to build the Berlin Wall. This section, preserved as a monument, shows how the concrete is held together by steel bars.

Calcium in water

Water from the faucet is called fresh water, but it is very rarely pure water. Besides tiny, generally harmless microbes, there are almost always some minerals dissolved in it, the most important being calcium and magnesium compounds. It is these compounds that make water "hard." If water contains more than 120 milligrams per liter of dissolved calcium and magnesium compounds, it is said to be "hard." People living in areas where the water is drawn from limestone rocks often have hard water. Hard water makes it difficult to create a lather with soap, and the compounds—especially calcium carbonate—are often deposited in a white crust, called limescale, around hot-water faucets and inside electric kettles.

Hard news

Water containing calcium bicarbonate (calcium hydrogen carbonate) is said to be "temporarily hard," because calcium bicarbonate can be removed by boiling, to leave behind a deposit of carbonate, or limescale. This is often visible as a white rim around bathtubs, a dull film coating faucets, and as a "fur," or crust, on the elements in kettles and water heaters.

Water containing calcium and magnesium salts other than bicarbonate is said to be "permanently hard," because no amount of boiling removes them from the water. They do not leave a visible scale, but they do make it hard for soap to lather.

Hard water can cause deposits of calcium carbonate, called limescale, to build up on the heating elements of boilers and kettles, as shown.

Water softening

Water hardness does no harm, but many people find limescale around water heaters and in kettles ugly or annoying because it reduces heating efficiency. It can also block up pipes. Different types of chemical product have been developed to combat scaling. Most are based on acid, which dissolves the alkaline calcium carbonate. Simple acids such as vinegar (acetic acid) and lemon juice (citric acid) remain among the most effective—and harmless—of descalers. Many people also want to soften hard water because it leaves a scum on dishes and makes it hard to get a good lather with detergents or soap powder.

Because some people feel that hard water leaves clothes feeling slightly crusty after washing, soap manufacturers may add the element phosphorus to soaps and detergents. This softens the water and makes soap work better, but it also means that huge doses of phosphates pour into rivers when washing-machine water drains away. Phosphates act as fertilizers and can make algae in the water grow so quickly that they choke all other life—a problem called "algal bloom." That is why today many detergents and soaps are advertised as phosphate-free.

Another approach is to feed the water supply through a filter inside the tank. The filter contains clay or resin coated in table salt. This absorbs calcium and magnesium minerals, as calcium and magnesium ions are swapped for sodium ions from the salt.

SEE FOR YOURSELF

MAKING AND REMOVING LIMESCALE
Limescale is usually seen as a nuisance as it often forms an unsightly coating or film on containers in which water is regularly heated—but it is easy to make, and easy to remove.

Making limescale
● Pour a little water into a heatproof glass dish or a clean nonstick saucepan.
● Ask an adult to gently boil the water until it has all boiled away. As the water evaporates, the solids dissolved in the water become visible.
● When the water has completely evaporated, a solid white film will be left around the side of the dish or saucepan. The film is calcium carbonate, which has been deposited as the molecules of liquid water changed into gas molecules and escaped into the atmosphere.

Removing limescale
● To remove the limescale, squeeze the juice from a lemon, or take some bottled lemon juice.
● Drip the lemon juice on to the limescale.
● The calcium carbonate dissolves as it reacts with the mild acid (called citrid acid) in the lemon juice. The calcium carbonate is changed into calcium bicarbonate (calcium hydrogen carbonate). Calcium bicarbonate is far more soluble in liquids than calcium carbonate, so the calcium bicarbonate disappears into solution.

This landscape is characterized by limestone that has become cracked and fissured in the form of a "pavement." The water that flows through these rocks has above-average amounts of calcium dissolved in it and is classified as "hard."

Calcium in industry and agriculture

Calcium compounds have many uses, both in industry and agriculture.

Lime on the land

For centuries, farmers and gardeners have used calcium compounds to make soil more fertile. Calcium compounds improve soil condition by binding grains of soil into bigger clumps so that clay soils drain better and do not become waterlogged. They can also neutralize excess acidity.

The calcium compound most widely used in agriculture is slaked lime (calcium hydroxide). This improves soil in two ways. First, it reduces acidity as the hydroxide ions bind with excess hydrogen ions in the soil to form water. Second, positively charged calcium ions help to bind soil particles by sticking to the surface of negatively charged clay particles in soil.

If the soil is already neutral, slaked lime could make it too alkaline for many plants. In these soils, crushed limestone is better, since it remains undissolved and inactive until the soil acidity increases sufficiently to dissolve it.

Lime in industry

Calcium compounds are used in almost every branch of industry: for example, chalk to make false teeth, putty, glass, and paper; and limestone to make sodium carbonate (washing soda), glass, steel, cement—and quicklime, of course. In steelmaking, limestone removes silicon-based impurities in the blast-furnace. Quicklime is used to speed up drying in

Calcium hydroxide is widely used by farmers to reduce the acidity of soils that are too acidic for the crops they want to grow.

This factory worker is using calcium compounds to make glass.

building work and in making bricks. Calcium sulfate (from gypsum) is used in everything from plasterboard and plaster casts to writing and drawing chalks. Slaked lime is used to remove acidic gases in various industrial processes and to make industrial bleaching powders.

Breath detector

Limewater, the solution made when slaked lime is dissolved in water, is used to detect the presence of carbon dioxide. If gas containing carbon dioxide is bubbled through limewater, the solution turns milky as the carbon dioxide reacts with it to form calcium carbonate, which is insoluble in water. If more carbon dioxide is bubbled through, the solution will clear again as more of the calcium carbonate is turned into soluble calcium bicarbonate (calcium hydrogen carbonate).

ATOMS AT WORK

Limewater is calcium hydroxide dissolved in water. It has one calcium ion attached to two hydroxide ions. When calcium hydroxide reacts with carbon dioxide, the calcium hydroxide joins with the carbon atom and one of the oxygen atoms from the carbon dioxide, forming calcium carbonate. The other oxygen atom from the carbon dioxide joins with the hydrogen from the hydroxide ions to form water.

Oxygen

Hydrogen

Calcium

Carbon

Calcium hydroxide
$Ca(OH)_2$

Carbon dioxide
CO_2

Calcium carbonate
$CaCO_3$

Water
H_2O

The reaction that takes place between limewater and carbon dioxide can be written like this:

$$Ca(OH)_2 + CO_2 \rightarrow CaCO_3 + H_2O$$

This shows that a molecule of calcium hydroxide reacts with a molecule of carbon dioxide to form one molecule of calcium carbonate and one molecule of water.

Calcium in the body

Calcium is one of the most important building blocks of life. Although it makes up only 2 percent of your body weight, it is an essential ingredient and serves many vital functions.

Bones

Bones grow like every other body cell. But in time, some cells become surrounded by hardened mineral deposits. It is these hardened minerals that give bones their rigidity and so support the body. The dominant mineral is calcium phosphate, but the harder outer walls of bones and the rods that carry the blood vessels through the soft central bone marrow are made of calcium carbonate. Although these minerals are lifeless, bones themselves are living tissue.

Bone contains lots of calcium, so children need plenty of calcium in their diet. That is one reason why they are encouraged to drink milk, which contains lots of calcium. Calcium deficiency, perhaps caused by malnutrition, can lead to weak or deformed bones.

Although calcium is always essential for good health, our need for calcium reduces as we get older. This is because the bones are simply being renewed, not grown. However, some older people suffer from osteoporosis, which makes the bones hollow and easy to break through lack of calcium. This can be part of the normal changes we experience as we grow older, but it can also be caused by

Without calcium compounds, the body could have no skeleton to support it and protect its soft internal tissues.

CALCIUM IONS HELP PUMP IRON

There are positively charged calcium ions inside every cell in the body. These ions, along with a few similar ions of potassium, sodium, magnesium, chloride, and phosphate, are crucial to the functioning of body cells, helping them to keep the correct chemical balance. Without these ions, the larger, more complex molecules such as proteins would probably fall apart.

Calcium ions are also vital for various chemical reactions and are directly involved in a number of important tasks, such as the transmission of nerve signals and the working of muscles. Crucial hormones and chemical messengers in the body cells, such as cyclic AMP, depend entirely on calcium ions for their working. Experiments show that an injection of even a small number of calcium ions into a muscle will make it contract.

hormonal changes in women who reach menopause (when they pass beyond childbearing age), which is why osteoporosis is sometimes treated with the female hormone estrogen.

Teeth

Teeth are coated with enamel, which is an especially hard form of calcium carbonate called aragonite. Enamel makes teeth very tough and resistant to wear. This is just as well because your second, or mature, set of teeth, once grown, do not grow again and have to be repaired by dentists if broken.

Like all forms of calcium carbonate, tooth enamel is attacked by acids. This is why teeth are decayed by the acids that are created in the mouth when bacteria digest sweet foods.

This woman suffers from osteoporosis, a softening of the bones that occurs in some elderly people due to calcium deficiency.

Periodic table

Everything in the universe is made from combinations of substances called elements. Elements are the building blocks of matter. They are made of tiny atoms, which are too small to see.

The character of an atom depends on how many even tinier particles called protons there are in its center, or nucleus. An element's atomic number is the same as the number of protons.

Scientists have found around 110 different elements. About 90 elements occur naturally on Earth. The rest have been made in experiments.

All these elements are set out on a chart called the periodic table. This lists all the elements in order according to their atomic number.

The elements at the left of the table are metals. Those at the right are nonmetals. Between the metals and the nonmetals are the metalloids, which sometimes act like metals and sometimes like nonmetals.

○ On the left of the table are the alkali metals. These elements have just one electron in their outer shells.

○ On the right of the periodic table are the noble gases. These elements have full outer shells.

○ Elements in the same group have the same number of electrons in their outer shells.

○ Elements get more reactive as you go down a group.

○ The number of electrons orbiting the nucleus increases down each group.

○ The transition metals are in the middle of the table, between Groups II and III.

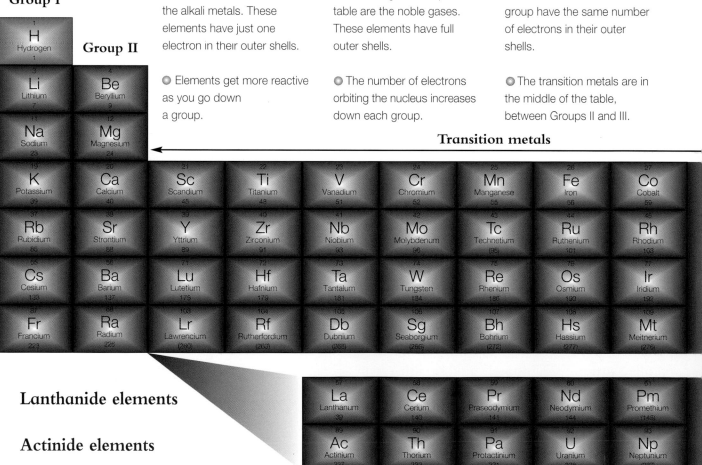

Group I

Group II

Transition metals

Lanthanide elements

Actinide elements

The horizontal rows are called periods. As you go across a period, the atomic number increases by one from each element to the next. The vertical columns are called groups. Elements get heavier as you go down a group. All the elements in a group have the same number of electrons in their outer shells. This means they react in similar ways.

The transition metals fall between Groups II and III. Their electron shells fill up in an unusual way. The lanthanide elements and the actinide elements are set apart from the main table to make it easier to read. All the lanthanide elements and the actinide elements are quite rare.

Calcium in the table

Calcium has atomic number 20, so it has 20 protons in its nucleus. It is one of six alkaline-earth metals in Group II of the table, which means it has two electrons in its outermost shell.

Calcium is highly reactive and forms a large number of compounds, many of them naturally occurring.

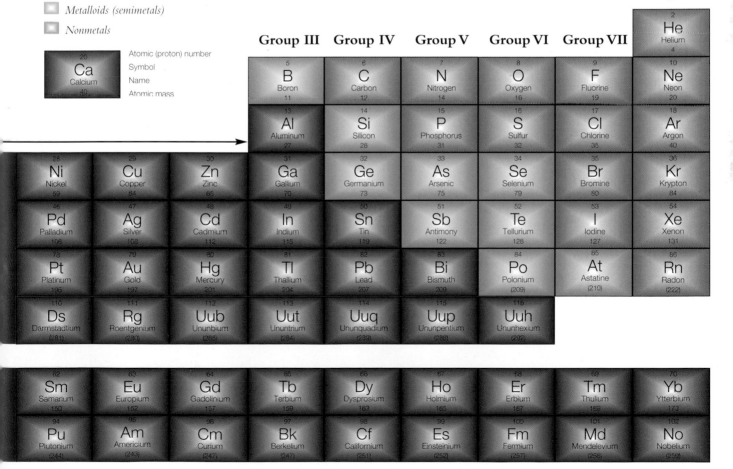

Chemical reactions

Chemical reactions are going on around us all the time. Some reactions involve just two substances; others many more. But whenever a reaction takes place, at least one substance is changed.

In a chemical reaction, the atoms stay the same. But they join up in different combinations to form new molecules.

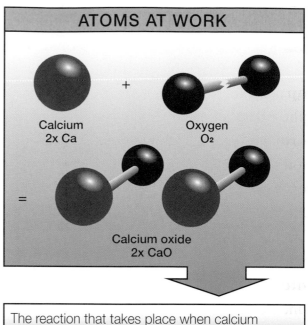

ATOMS AT WORK

+

Calcium
2x Ca

Oxygen
O_2

=

Calcium oxide
2x CaO

The reaction that takes place when calcium reacts with oxygen, can be written like this:

$$2Ca + O_2 \rightarrow 2CaO$$

Calcium is highly reactive and forms many compounds, among them calcium oxide. Calcium oxide is the basic raw material of cement—seen here being unloaded as freight.

Writing an equation

Chemical reactions can be described by writing down the atoms and molecules before and the atoms and molecules after. Since the atoms stay the same, the number of atoms before will be the same as the number of atoms after. Chemists write the reaction as an equation. This shows what happens in the chemical reaction.

Making it balance

When the numbers of each atom on both sides of the equation are equal, the equation is balanced. If the numbers are not equal, something is wrong. So the chemist adjusts the number of atoms involved until the equation does balance.

Glossary

apatite: A mineral form of calcium phosphate.

aragonite: A form of calcium carbonate found in seashells and tooth enamel.

atom: The smallest part of an element having all the properties of that element. Each atom is less than a millionth of an inch in diameter.

atomic mass: The number of protons and neutrons in an atom.

atomic number: The number of protons in an atom.

bond: The attraction between two atoms, or ions, that holds them together.

calcite: A crystalline form of calcium carbonate.

chalk: A form of calcium carbonate, produced from plant and animal remains.

compound: A substance made of two or more elements chemically joined together.

corrosion: The eating away of a material by reaction with other chemicals, often oxygen and moisture in the air.

descaler: A substance used to prevent limescale build-up in kettles and water heaters.

dolomite: A mineral form of calcium—calcium magnesium carbonate.

electrode: A material through which an electrical current flows into, or out of, a liquid electrolyte.

electrolysis: The use of electricity to change a substance chemically.

electrolyte: A liquid that electricity can flow through.

element: A substance that is made from only one type of atom. Calcium is one of the elements called an alkaline-earth.

gypsum: A mineral form of calcium sulfate, used to make plaster.

ion: A particle of an element similar to an atom but carrying an additional negative or positive electrical charge.

lichen: A moss-like plant growth that often colonizes limestone buildings, walls, and headstones.

limestone: A naturally occurring mineral form of calcium carbonate, mostly used in the building industry.

metal: An element on the left-hand side of the periodic table.

oolite: A form of limestone that contains tiny balls of calcite.

periodic table: A chart of all the chemical elements laid out in order of their atomic number.

precipitate: A solid substance that drops out of a solution.

quicklime: A name for calcium oxide.

slaked lime: Another name for calcium hydroxide.

spar: A name given to some forms of calcite, such as dogtooth and Iceland spar.

travertine: A form of calcite found in association with thermal springs.

Index